Eric Holtschke

The Role of Media during the Arab Spring with Particular Focus on Libya

Der GRIN Verlag publiziert seit 1998 wissenschaftliche Arbeiten von Studenten, Hochschullehrern und anderen Akademikern als eBook und gedrucktes Buch. Die Verlagswebsite www.grin.com ist die ideale Plattform zur Veröffentlichung von Hausarbeiten, Abschlussarbeiten, wissenschaftlichen Aufsätzen, Dissertationen und Fachbüchern.

Eric Holtschke

The Role of Media during the Arab Spring with Particular Focus on Libya

GRIN Verlag

1. Auflage 2013
Copyright © 2013 GRIN Verlag GmbH
http://www.grin.com
Druck und Bindung: Books on Demand GmbH, Norderstedt Germany
ISBN 978-3-656-38006-1

Charles University in Prague

Faculty of Social Sciences

Geopolitical Studies

Course: Geopolitics of the Middle East

Winter Semester 2012/2013

The Role of Media during the Arab Spring with Particular Focus on Libya

Submitted by:

Eric Holtschke

Course of Studies:

Master's Degree Programme Geopolitical Studies

1st Study Semester

Table of Contents

The Role of Media during the Arab Spring with Particular Focus on Libya

by Eric Holtschke

1. Introduction

When the people in the Arab world started to go out on the streets in December 2010 for the very first time for years, no one of the demonstrators and protesters who were fighting for their rights to live in a better world, in peace, freedom and democracy expected that level of change as a result of their behaviour. Those inconceivable aftermaths did not exist neither in the minds of the people in Tunisia, the nucleus of the Arab Spring, nor in the minds of the people who were joining that wave of protest country by country afterwards. Before that uprising arrived Libya, which is in the focus of attention here, in February 2011, the revolt expanded over the Middle East and North Africa. Never before mass media, including radio, television and, especially the internet (with Facebook, Twitter and YouTube), was taken such a significant role during uprisings since the end of the year 2010. No one of the participants, who were involved in this uprising, no one of the politicians, no one of the citizens in the Western world, no one of the former leader of those regimes and no one of the media companies could count on this great measure of transformation in the Arab world. States were dissolved, former inhuman leaders of a regime were captured or killed and new fundaments of a modern, democratic and liberalized nation were created by the people or the opposition of those long lasting dictatorial and authoritarian leaders.

Taking a deeper look to the status of mass media during and after the Arab Spring it can be mentioned that particularly newspapers, as a component of mass media – which are currently in a critical financial situation all over the world – were presenting a kind of positive picture of the role of media regarding to the protest wave. Thus, Ramesh Srinivasan, an Associate Professor in Media/Information Studies at University of California, Los Angeles, published on the web page of "Al Jazeera" the following headline: *"Taking power through technology in the Arab Spring"*[1]. Further he said that *"social media is no longer the domain of the liberal youth, empowering different agendas across the political map"*, and he added that in the *"[...] Arab world, the protests and riots would not have occurred without YouTube [...]"*[2]. On the platform "Policymic" is written: *"Twitter Revolution: How the Arab Spring*

---

[1] Al Jazeera: Taking power through technology in the Arab Spring, viewed on
http://www.aljazeera.com/indepth/opinion/2012/09/2012919115344299848.html, retrieved on 24th of January 2013.
[2] Ib., retrieved on 24th of January 2013.

was helped by Social Media."[3] And it is explained further: *"Ultimately, public information supplied by social networking websites has played an important role during modern-day activism, specifically as it pertains to the Arab Spring."*[4] The "Small Wars Journal" on the other hand has a more negative point of view about the importance of social media during Arab Spring, as it is described as follows: *"Social media eliminates two important impediments to communication – distance and delay – creating new pathways for people to connect in near-real time. But the social networks themselves are insufficient to drive a population to the streets."*[5]

The fact is, that activists being significantly involved in the Arab Awakening were using social media networks as an important tool to support the spreading of the protest wave. Because: *"Social media, perhaps thanks to the international and domestic hype, has a cache [...] that it did not have before the events of 2011."*[6]

1.1 Problem Statement

The intention of this present paper is to contribute an insight into a very specific topic of the interdependence between the role of mass media and the course and aftermaths of the uprisings in die Arab world. Related to the limited number of pages the paper mainly deals with focus on the Arab Spring in Libya and its connection and dependency to the social mass media in the period of time from the beginning of the spreading of the demonstrations in the Arab world over Libya until nowadays.

That paper was realized on the basis of a scientific literature and internet research, verified by numerous quotations from newspapers, political platforms, internet articles and specialised books and content-related movies. The target of this present paper is to close a gap in an area little explored by academics until now.

The central question of that paper is: Did media affect the course of the Arab uprising? If yes, how has media, e.g. social networks like Facebook and Twitter as well as mass media like television and radio, affected the protest wave and its spreading in the Arab Spring and the outcomes of that protests in Libya? It gives rise to the following sub-questions: How can

---

[3] Twitter Revolution. How the Arab Spring was helped by Social Media, viewed on http://www.policymic.com/articles/10642/twitter-revolution-how-the-arab-spring-was-helped-by-social-media, retrieved on 24th of January 2013.

[4] Ib., retrieved on 24th of January 2013.

[5] Small Wars Journal: Social Media and the Arab Spring, viewed on http://smallwarsjournal.com/jrnl/art/social-media-and-the-arab-spring, retrieved on 24th of January 2013.

[6] Al Jazeera: Taking power through technology in the Arab Spring, retrieved on 24th of January 2013.

the course of Arab Awakening can be described as a scenario without the existence of global media? And: Have media supported the Arab Spring or, vice versa, the Arab Spring media?

1.2 Structure

In order to be able to analyse and to evaluate the issue regarding to the questions mentioned above, necessary terminologies which are in the focus of this discussion need to be specified, contents-related limited and finally defined at the beginning of the main section of this paper. Those nomenclatures are basically required for a better understanding of this present scientific paper. They can be seen as the fundament to a profound answering of the research questions. To maintain coherence of this paper the issue the thesis is dealing with will be presented in a primarily chronological and secondarily in a thematic order.

The approach to this topic will be done in the first content-related point 3: "The awakening of the Arab world: The Arab Spring in Libya". This chapter is focussed on the backgrounds, the course and the outcomes of the Arab Spring with concentration on Libya. Regarding to the backgrounds the institutional frameworks of transitioning Libya is explained here. A indispensable historical abstract of the emergence, the development and the results of the uprisings in the Arab World in consideration of Libya can be found, too. The results deal about economical, political and societal aftermaths. This topic is rounded off by some interesting and unexpected statistics, numbers and quotations.

The second content-related chapter mainly deals deeply about the relation, the connection and the interdependency of media and the protests during the Arab Spring. Particularly the role of social mass media regarding to the course of the Arab Awakening is discussed, taking particular account to the research question mentioned in the introduction. Thus, current articles in newspapers or in the internet as well as two important moves which can be seen as a fundament to write about this issue are in the foreground here. They have been analyzed in order to achieve a possible scenario, mentioned in point 4.3. The scenario is including a specifically designed imaginary course of Arab Spring in the case of non-existence of mass media or the cut off of the web connection in the countries affected by the Arab Spring, especially Libya, based on resources which are dealing about that issue. Aspect 4.2 is leaving the main topic slowly, being concentrated on the target and on the boundaries of the functions of mass media in general. It can be recognized as a summary of its task with regard to contents.

Before this present paper will be finished by a detailed bibliography, it ends with a final abstract in point 5 of the table of contents. The abstract is focussed on a summarisation regarding to the aspects the paper was dealing with. Furthermore, the writer of this paper want to provide the feasibility to construct an optional course how media will be created in the near future, especially in consideration of its ability and potential in the transition of dictatorial regimes, in the spread of protest waves or in the democratization of nations.

Direct quotations and original used text extracts are presented in quotation marks and in a cursive style to accentuate its importance. It is deployed to simplify the flow of reading. Non-cursive words, phrases, sentences or passages of sentences are either colloquial used terms or expression from which the author would consciously like to distance himself.

1.3 State of Research

The state of research has to be evaluated as literarily weak due to a relatively new research area, which is mainly discussed in the scientific field of media, film and information studies as well as in communication science and political science, more or less in peace research and in global and regional studies. An important role is playing by the media itself, in particular newspapers and the internet. Written books about this issue are still quite rare.

The most important representatives of written literature in consideration of media and Arab Spring are Wael Ghonim and Lina Ben Mhenni. Ben Mhenni is working as a university lecturer at the University of Tunis. She describes herself as an internet activist and political blogger, fighting for human rights and against censorship. Her book, published 2011 in French in a German publishing house, is called "Tunisian Girl – Blogueuse pour un printemps arabe"[7]. It can simply translated as "Get connected".

A second so called internet activist is Ghonim. Born in Cairo he was voted as one of the most influential personalities by the American "Time Magazine". He can be described as a key figure during the revolution in Egypt, nowadays he is well-known around the world. His book, published last year, entitled "Revolution 2.0"[8], includes the following sub-headline: "The power of the people is stronger than the people in power."

Both, Ben Mhenni and Ghonim, used their power gained during the protest wave to publish a book and to discuss this issue in the publicity. But they are not the only one who were dealing about that.

---

[7] Lina Ben Mhenni: Vernetzt euch!, Berlin 2011.
[8] Wael Ghonim: Revolution 2.0. The Power of the People Is Greater Than the People in Power. A Memoir, New York 2012.

The state of research within printed articles needs to be evaluated as comprehensive. In the scientific field Ramesh Srinivasan[9], an Associate Professor in Media/Information Studies at University of California in Los Angeles, who was already mentioned in the introduction, is playing a leading role. He published and he is still publishing countless numbers of articles, usually in the internet, in magazines or newspapers.

A second nucleus can be found in different institutions, think tanks, universities and governmental or non-governmental organisations around the globe. One of the most important was written by Echo Keif and published by George Mason University, Washington D.C.: "We Are All Khaled Said: Revolution and the Role of Social Media."[10] The German Institute for Foreign Relations[11], placed in Stuttgart, has to be mentioned here, as well as the International Association for Media and Communication Research[12], the Association for Progressive Communications (APC)[13] and "Freedom House". The NATO was also dealing intensively about this topic.

2. Definitions

2.1 Arab Spring

The term "Arab Spring" is defined by the online encyclopaedia "Britannica" as follows: *"Wave of pro-democracy protests and uprisings that took place in the Middle East and North Africa beginning in 2010 and 2011, challenging some of the region's entrenched authoritarian regimes. Demonstrators expressing political and economic grievances faced violent crackdowns by their countries' security forces."*[14] The protest movement, started in Tunisia and Egypt, was called "Jasmine Revolution" by the media. The movement spread over the countries quickly and overwhelmed the security forces in the Arab world. *"The effects of the Arab Spring movement were felt elsewhere throughout the Middle East and North Africa as many of the countries in the region experienced at least minor pro-democracy protests. In Algeria, Jordan, Morocco, and Oman, rulers offered a variety of concessions,*

---

[9] Al Jazeera: Taking power through technology in the Arab Spring, retrieved on 25th of January 2013.

[10] Echo Keif: We Are All Khaled Said. Revolution and the Role of Social Media, Fairfax 2011.

[11] Anja Türkan: Digital Diplomacy. Der Wandel der Außenpolitik im digitalen Zeitalter, Stuttgart 2012.

[12] Annabelle Sreberny: New Media and the Middle East. Thinking allowed, in: University of Michigan: Journal, No. 2 (Spring 2012), viewed on: http://www.lsa.umich.edu/UMICH/ii/Home/II%20Journal/Documents/2012 spring_iijournal_article1_sreberny.pdf, retrieved on 29th of January 2013.

[13] Alex Comninos: Twitter revolutions and cyber crackdowns. User-generated content and social networking in the Arab spring and beyond, 2011.

[14] Encyclopædia Britannica Online: Arab Spring, viewed on http://www.britannica.com/EBchecked/topic/1784922/Arab-Spring, retrieved on 26th of January 2013.

ranging from the dismissal of unpopular officials to constitutional changes, in order to head off the spread of protest movements in their countries."[15]

2.2 Media

The term "media" in general can be defined as an *"umbrella term covering all technological means which are used for circulation of information, especially for communicative devices"*[16]. Newspapers, magazines, books, posters, radio, television, movies as well as the internet can be count to the so called "communicative devices". These examples are, together with e-mails, internet blogs, Facebook, Twitter and YouTube, a part of the global mass media. They are *"usually less influential than the social environment, but they are still significant, especially in affirming attitudes and opinions that are already established"*[17]. Mass media can *"reinforce latent attitudes and 'activate' them, prompting people to take action"*[18]. It also plays another important role by giving individuals the possibility to know what other people in their social environment think and by providing political leaders a large amount of followers. *"In this way the media make it possible for public opinion to encompass large numbers of individuals and wide geographic areas. It appears, in fact, that in some European countries the growth of broadcasting, especially television, affected the operation of the parliamentary system."*[19] In consideration of the discussed issue of Libya it has to be mentioned that in areas, in those mass media are not well developed, e.g. in developing countries or in countries where the media are under control of political leaders, the oral communication is usually able to perform almost the same functions as broadcasting and press, even though slower and content-related limited. *"In developing countries, it is common for those who are literate to read from newspapers to those who are not, or for large numbers of persons to gather around the village radio or a community television."*[20]

A further part of mass media can be recognized in news media. It is focussing on *"the public's attention on certain personalities and issues, leading many people to form opinions about them"*[21]. Facebook, Twitter, YouTube and other internet portals are a part of social media or social network. A "social network" is a *"online community of individuals who*

---

[15] Ib., retrieved on 26th of January 2013.

[16] Bibliographisches Institut & F.A. Brockhaus AG: Medien, viewed on http://www.brockhausenzyklopaedie.de/be21_article.php?document_id=0x0919d990@be, retrieved on 26th of January 2013.

[17] Encyclopædia Britannica Online: Public Opinion. The mass media, viewed on http://www.britannica.com/EBchecked/topic/482436/public-opinion/258760/The-mass-media#ref397897, retrieved on 26th of January 2013.

[18] Ib., retrieved on 26th of January 2013.

[19] Ib., retrieved on 26th of January 2013.

[20] Ib., retrieved on 26th of January 2013.

[21] Ib., retrieved on 26th of January 2013.

exchange messages, share information, and, in some cases, cooperate on joint activities."[22] Further it is written: *"Eschewing the anonymity that had previously been typical of the online experience, millions of people have flocked to social networking sites where members create and maintain personal profiles that they link with those of other members. The resulting network of 'friends' or 'contacts' who have similar interests, business goals, or academic courses has replaced for many people, especially youth, older concepts of community."*[23]

All these aspects can be summarized under one common term: communication. It is simply described as *"the exchange of meanings between individuals through a common system of symbols"*[24]. Ivor Armstrong Richard, a British writer and literary critic, defined "communication" in the year 1928 in a timeless manner: *"Communication takes place when one mind so acts upon its environment that another mind is influenced, and in that other mind an experience occurs which is like the experience in the first mind, and is caused in part by that experience."*[25]

3. The Awakening of the Arab world: The Arab Spring in Lybia
3.1 Backgrounds

The emergence of the Arab Spring started on 17th of December 2010 in Tunisia. The wave of protests expanded quickly over the Middle East and North Africa (Maghreb). It resulted in the escaping or resignation of the political leaders in power, Zine el-Abidine Ben Ali in Tunisia in January 2011 and Husni Mubarak in Egypt in February 2011. The Arab Spring has affected two states within a couple of weeks which seemed to be the most stable in that region: *"The Tunisian government attempted to end the unrest by using violence against street demonstrations and by offering political and economic concessions."*[26] The government of Egypt acted similar. It tried to control the uprisings by providing acknowledgements, but the government failed. After a 30 years lasting reign of terror Mubarak lost support by the military.

In December 2010 the uprising in Tunisia reached Algeria and spread over the country in the beginning of the next month. On 7th of January, Jordan has been affected by the Arab Spring, in the end of the month the protest wave reached Saudi Arabia and Yemen.

---

[22] Encyclopædia Britannica Online: Social Network, viewed on http://www.britannica.com/EBchecked/topic/1335211/social-network, retrieved on 26th of January 2013.
[23] Ib., retrieved on 26th of January 2013.
[24] Here and follows: Encyclopædia Britannica Online: Communication, viewed on http://www.britannica.com/EBchecked/topic/129024/communication, retrieved on 26th of January 2013.
[25] Ib., retrieved on 26th of January 2013.
[26] Here and follows: Encyclopædia Britannica Online: Arab Spring, retrieved on 27th of January 2013.

Demonstrations could also be found in Mauritania within the first two months of 2011. Afterwards, the protest wave spread over Palestine, Bahrain, Libya, Djibouti, Kuwait, Oman, Morocco, Sudan and Iraq in February 2011. It came to an bloody act of violence and to violation between the opposition fighting on the streets for a changing state and the dictatorial regimes fighting for staying in power. Syria has been affected within that domino effect as one of the last countries in mid-March 2011.

Illustration 1[27]: Countries and their political status being affected by the Arab Spring

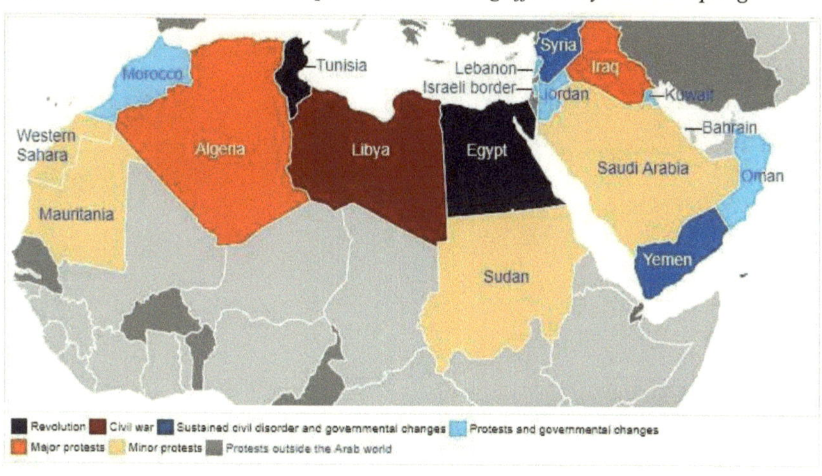

The Arab Spring spread over Libya on 15th of February 2011 for various reasons. Firstly, the uprisings in the Arab world spread in the way of a domino effect over neighbouring states. Secondly, people living there were dissatisfied with the political, social and economic situation and circumstances to a large extent. Especially displeasure regarding to the regime and its security apparatus, lacking of democratic culture and participation, a high unemployment rate, in particular of the youth, as well as an increasing of food and energy prices connected with a fast growth of population in Libya. Thirdly, all classes of society were affected by an extensive corruption in all spheres of policy, economy and administration. Fourthly, the former leader Muammar al-Gaddafi, born on 19th of June 1942, died on 20th of October 2011, were ruling Libya like a dictator since 1969. He presented himself in publicity as a flamboyant personality. Fifthly, the role of social mass media and the spreading of information all around the globe in a couple of seconds were supporting the

---

[27] Countries and their political status being affected by the Arab Spring, viewed on https://moodle.hampshire. edu/pluginfile.php/69048/course/section/31642/arab-spring-map1.jpg, retrieved on 31st of January 2013.

spreading of protests. Sixthly, the spirit of the uprisings could be found in every social class: left and right-winged people were as well involved as liberals, conservatives, Muslims and Christians. Seventhly, the protest wave were simply popular within the dissatisfied citizens. And eighthly, *"antigovernment rallies were held in Banghāzī by protesters angered by the arrest of a human rights lawyer, Fethi Tarbel"*[28].

3.2 Course

After the Arab Spring spread over Libya in the mid of February 2011, the demonstrators were calling for an amnesty of political prisoners. They were calling for a resignation of al-Gaddafi, too. *"Libyan security forces used water cannons and rubber bullets against the crowds [...], a pro-government rally orchestrated by the Libyan authorities was broadcast on state television."*[29] The protesters intensified their uprisings, the government were striking back with targeted violence, using an extremely harsh. *"Demonstrators also were attacked with tanks and artillery and from the air with warplanes and helicopter gunships."*[30] And: *"Security forces and squads of mercenaries fired live ammunition into crowds of demonstrators."*[31] The regime shut down internet and mobile phone connections and prohibited digital communication in general. On 21st of February 2011, Sayf al-Islam, one of Gaddafi's sons, cautioned against a civil war in the television which might be possible if the demonstrations would not be stopped. Step by step Gaddafi were losing power: the Libyan slowly armed forces sided with the protests. *"As demonstrators acquired weapons from government arms depots and joined forces with defected military units, the anti-Qaddafi movement began to take the form of an armed rebellion."*[32]

The protest wave went into a civil war shortly afterwards and ended officially on 23rd of October 2011. The civil war led to a horrible amount of deaths and casualties: 30,000 people were killed and 60,000 people were injured. At least, the civil war, which can be basically explained as a war of a rebel group challenging the government merely within a state with different types of group violence, e.g. riots, protests, pogroms and coups, resulted in the overthrow of Gaddafi's regime. The further developments of the political sphere led to a division of the new political leadership. Resulting from those differences a so-called National

---

[28] Here and follows: Encyclopædia Britannica Online: Libya Revolt of 2011, viewed on http://www.britannica.com/EBchecked/topic/1766291/Libya-Revolt-of-2011, retrieved on 28th of January 2013.
[29] Ib., retrieved on 28th of January 2013.
[30] Ib., retrieved on 28th of January 2013.
[31] Ib., retrieved on 28th of January 2013.
[32] Ib., retrieved on 28th of January 2013.

Transitional Council was instituted which controlled the Eastern part of the country. After all, the civil war can be categorized as a civil war of the third category: serious political violence[33]. The objectives within such a civil war are usually focussed on strategic authority, including control of human and of material resources. The population feels able to respond regarding to changing conditions of specific localized operations. The effects of serious political violence are uneven distributed, mainly targeting militias, leaders and symbolic targets.

Libya went straight into an civil war because almost the half of the political authorities of the former regime defected to the side of the opposition, al-Gaddafi were on the step to lose his power. The Security Council of the United States imposed sanctions against Gaddafi, for example a ban of travelling of members of the regime as well as an arms embargo and freezing of foreign bank accounts. An empowerment to intervene in Libya with military measures to protect the local civilians was given under authority of the United Nations. The United Nations Security Council adopted the Resolution 1973 on 17th of March 2011. Two days later, the United States of America, Great Britain and France started an air-see blockade as well as a raid on military facilities and on armed forces of Gaddafi's regime. The involved states also imposed a no-fly zone over Libya. The main target of the military intervention was to support the oppositional armed forces in the occupation of cities in the Western part of the country. On 27th of March 2011 the North Atlantic Treaty Organization (NATO) started to lead the whole military intervention. An earlier intervention was prevented by Turkey before. The Minister of Foreign Affairs of Libya, Moussa Koussa, escaped to Great Britain on 30th of March 2011. It was recognized as a loosing of power of Gaddafi's armed forces.

The fights continued for months. *"In August 2011 rebel forces advanced to the outskirts of Tripoli, taking control of strategic areas, including the city of Zāwiyah, the site of one of Libya's largest oil refineries."*[34] In the end of August 2011 oppositional armed forces captured the capital city Tripoli; the National Transitional Council, which emerged as a rebel leadership council, formed by local rebel groups, was moved to the capital afterwards. The council's targets were to *"act as the rebellion's military leadership"*, to be the *"representative of the Libyan opposition"*, to *"provide services in rebel-held areas"* and to *"guide the country's transition to democratic government"*.[35] Basically the clashes were held in the periphery of Tripoli and in the region at the Gulf of Sidra.

---

[33] The ten categories of civil wars ideal-typically ranges from the first and weakest one (sporadic or expressive political violence) to the tenth and strongest one (extermination and annihilation).
[34] Encyclopædia Britannica Online: Libya Revolt of 2011, retrieved on 28th of January 2013.
[35] Ib., retrieved on 28th of January 2013.

On 20th of October 2011 Gaddafi's city of birth, Surt, were captured by the opposition, too. The leader of the regime was caught and killed there. On October 23rd 2011 the National Transitional Council declared Libya as a fully liberated country which can be seen as the official ending of the Arab uprising and the civil war in Libya.. Almost one month before, on 15th of September, the council achieved legitimacy by the international community. The United Nations General Assembly started to recognize the Nation Transitional Council as an official leader of Libyan government.

3.3 Results

Besides of imposition of sanctions by the international community against Gaddafi and his regime within the civil war, several outcomes need to be mentioned as a result of the Arab Spring and the civil war in Libya. After the ending of the civil war, the National Transitional Council could establish a working and functioning government in Libya to expand its power. The militias were disarmed which led to a conflict between different rebel groups until today. Furthermore, *"local rebel militias that had fought autonomously during the uprising, especially those in western Libya, were reluctant to submit to an interim government formed in eastern Libya with little input from the rest of the country [...]"*[36]. They were afraid of a possible connection of the council to former personalities of the regime. Militias still can be seen on Libya's territory today.

Illustration 2[37]: *Libya's freedom on the Internet 2011/2012*

	2011	2012	
INTERNET FREEDOM STATUS	n/a	Partly Free	POPULATION: 6.5 million
Obstacles to Access (0-25)	n/a	18	INTERNET PENETRATION 2011: 17 percent
Limits on Content (0-35)	n/a	9	WEB 2.0 APPLICATIONS BLOCKED: Yes
Violations of User Rights (0-40)	n/a	16	NOTABLE POLITICAL CENSORSHIP: No
Total (0-100)*	n/a	43	BLOGGERS/ICT USERS ARRESTED: Yes
* 0=most free, 100=least free			PRESS FREEDOM STATUS: Partly Free

The citizens of Libya are unsatisfied with the aftermaths of the revolution nowadays: the infrastructure is destroyed, ethnic and religious conflicts are still a common issue within

---

[36] Ib., retrieved on 28th of January 2013.

[37] Freedom House: Freedom on the Net. Libya 2012, viewed on http://www.freedomhouse.org/report/freedom-net/2012/libya, retrieved on 31st of January 2013.

the society and the economical structures have not been changed or merely slowly. People are still suffering from lacking democratic basics: no transparency, no freedom of press and media, violation of human rights, imprisoned citizens without a legal judgement, tortures in prisons, violence, missing of stable government as well as the humanitarian, medical, political and economical situation. In August 2012, the National Transitional Council moved into a so called General National Congress. Before, elections were held between 8th of June and 5th of July 2012. Mohammed Yusef el-Megarief became the new head of state and Mustafa Abu Schagur the head of government, but merely for a short period of time. After three weeks of being in office, he resigned. On 14th of October 2012, Ali Zeidan became the new Prime Minister of Libya.

Illustration 3[38]: Freedom on the Net 2012: Global Graphs

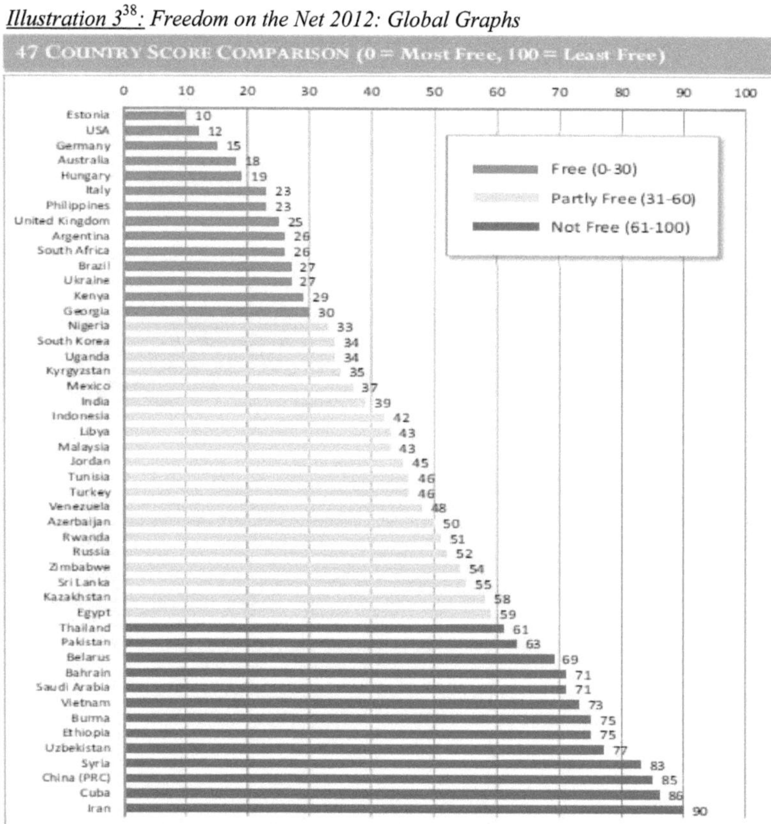

---

[38] Freedom House: Freedom on the Net 2012. Global Scores, viewed on http://www.freedomhouse.org/sites/default/files/resources/FOTN%202012%20-%20Tables%20and%20Charts%20FINAL.pdf, p. 4, retrieved on 31st of January 2013.

The freedom of press in Libya can be evaluated as partly free.[39] Within the last years the country improved from not free to a current partly free status *"due to significant improvements in media freedom and access to information"*[40]. Furthermore, it is said: *"Journalists are able to cover the news more freely than before, and with less threat of violence and intimidation."*[41]

The freedom on the net is partly free, too. Libya was positioned in the mid-range of the scale of freedom on the net in the previous year, as it is visible in the illustration above. The political changes after the death of al-Gaddafi *"were also reflected in the internet freedom landscape"*[42], is written in a report published by "Freedom House". Three important steps can be found in Libya regarding to the freedom on the net: *"[...] a highly restrictive environment under Qadhafi, a partial internet and telephone blackout for much of 2011, and a relatively open online information landscape since the rebel victory in October 2011."*[43] The using of social mass media after the revolution increased heavily in Libya. The accounts on Facebook have doubled to an amount of approximately 400,000 users. Facebook nowadays is the most visited webpage in this country due to be informed about the latest events going on. *"[...] Facebook, Twitter and other digital media have grown in popularity and been used to mobilize Libyans for activism around a variety of causes."*[44]

4. The Interdependency of Media and Arab Spring
4.1 Arab Spring without Media?

The emergence of the importance of the role of social mass media during the Arab Spring has to be recognized in Tunisia, when Mohamed Bouaziz, who was well known as a 26-year old salesman of fruits and vegetables on the street, light himself in the publicity. It was on December 17th when he protested against corruption within local political authorities and the police. Bouaziz did not survive his protest: he died on January 4th 2011. *„Soon, thousands of young people were protesting in the name of what happened in Sidi Bouzid and calling for the world to take notice. When a few news outlets did, they covered the protests not by sending in their own crews but by mining the Twitter hashtag (#Sidibouzid) and videos uploaded to*

---

[39] Freedom House: Freedom of the Press. Libya 2012, viewed on http://www.freedomhouse.org/report/freedom-press/2012/libya, retrieved on 31st of January 2013.
[40] Ib., retrieved on 31st of January 2013.
[41] Ib., retrieved on 31st of January 2013.
[42] Freedom House: Freedom on the Net. Libya 2012, retrieved on 31st of January 2013.
[43] Ib., retrieved on 31st of January 2013.
[44] Ib., retrieved on 31st of January 2013.

Facebook and YouTube from the ground."[45] After he died, the demonstrations were continued following his death across the internet and the country. *"Protesters uploaded videos of unrest whilst the authorities fought back with censorship and hacking"*[46], is said in a movie made by NATO. *"Social media played a role, but not a determinative role"*[47], Philip Seib, Director of Center of Public Diplomacy mentioned. As he has noted, the role of television as mass media was more important than the role of social mass media in the internet. The reasons for that can be found in the non-availability of the world wide web in the whole region. It differs heavily from developed and liberalized countries. 20 percent of the people in the Arab world are using the advantages of internet, only five percent of these are organized on Facebook. Television is reaching 80 percent of the citizens, especially in the bigger cities. *"And so the dominant medium was television, sometimes re-purposing material that it found on the social media sites."*[48]

In Egypt, Facebook played a more important role. A Facebook page existed which were calling for a public get-together on 25th of January 2011. The articulation of protesting was a result of the death of Khaled Said, a 28-year old man. *"Police officers had dragged him from an Internet café and beaten him to death"*[49], can be watched in the movie of NATO. 80,000 members of Facebook and more joined the pronunciamento. The Egyptian government stopped all internet and mobile phone connections afterwards for a couple of days, but the demonstrations continued. "We are all Khaled Said", the name of the Facebook group, has more than 300,000 likes.[50] *"Social media is much broader than sending 144 characters over Twitter or updating your status posts on Facebook"*[51], Robert McNab, Professor at Naval Postgraduate School in California, USA, said. And he continued in the interview: *"Those are useful, but if you look how people are sharing videos and information in Egypt, in Tunisia, in Libya, it is primarily through SMS, not through Twitter."*[52] Communication can not only be found in the internet. Mobile phones, text messages, telephones, radio and television are a part of mass media, too. This is resulting in a spread of dissemination of information. Philip Seib again was supporting the non-digital way of how to spread information regarding to the Arab

---

[45] NATO Review: The reconstruction will not be tweeted, viewed on http://www.nato.int/docu/review/2011/Social_Medias/Social_Media_changes/EN/index.htm, retrieved on 29th of January 2013.
[46] NATO Review: Arab Spring = Facebook revolution #1?, viewed on http://www.nato.int/docu/review/2011/Social_Medias/Arab_Spring/EN/index.htm, retrieved on 29th of January 2013.
[47] Ib., retrieved on 29th of January 2013.
[48] Ib., retrieved on 29th of January 2013.
[49] Ib., retrieved on 29th of January 2013.
[50] Facebook: We are all Khaled Said, viewed on http://www.facebook.com/elshaheeed.co.uk?fref=ts, retrieved on 29th of January 2013.
[51] NATO Review: Arab Spring = Facebook revolution #1?, retrieved on 29th of January 2013.
[52] Ib., retrieved on 29th of January 2013.
[52] Ib., retrieved on 29th of January 2013.

Spring: *"[...] for example, in Cairo people came out into the streets where they didn't even have television. [...] People were shouting into entryways of apartment houses: Come, join us. It was not any fancy technology."*[53]

The Arab Spring without media might was possible, because only one thing makes revolutions: the people. A well chosen example was the revolution against the communist regimes in the countries in central and East Europe in 1989 and afterwards. But media, especially the mass media like television, mobile phones and the internet were supporting the uprisings basically. Without media, the revolutions in the Arab World could be a longer-lasting combat witch hundred thousands of victims or more who were killed by the dictatorial leaders and armed forces of the regimes. Mass media shortened the Arab Springs in the advantage of the people who were involved. And (social) mass media have protected the life of thousands of people. It is simply safer to revolt in front of the notebook than in front of a regime's tank on the streets.

"It is evident that both new and old media have played significant and fascinating roles in the recent insurrections to topple autocratic regimes [...] New media can not be considered the epiphenomena of political movements but are rather significant tool of political mobilization."[54] But, and that is a fact: *"Insurrectionary movements have grown in countries with Internet penetration [...]."*[55] However, the Internet is used in Libya merely by six percent of the people, but it achieved strong impacts regarding to the mobilization of people. *"It is wrong to say that social networks can start a revolution. It is equally wrong to say that they did not play a significant role during the Arab Spring. Also, the use of social networking platforms and other Internet services for political struggle has not been invented during the Arab Spring, and it will not end with it."*[56]

The Arab Spring in Libya would not have happen in the way it did, because: *"Libyans in the diaspora used social media to spread the word and show support for the February 17 'day of anger' that launched the revolution. A large number of Libyans inside the country responded, changing their surname on Facebook to 'Libya' in a symbolic protest against Qadhafi's regime."*[57] In the beginning of the year 2011 64,000 active Twitter accounts and 200,000 Facebook accounts were counted in Libya. *"Once the uprising began, online media, blogs, and social networks played a visible role amplifying the voices of those inside the*

---

[53] Ib., retrieved on 29th of January 2013.

[54] Annabelle Sreberny: p. 3, retrieved on 29th of January 2013.

[55] Ib., p. 3 f., retrieved on 29th of January 2013.

[56] Markus Sabadello: The Role of New Media for the Democratization Processes in the Arab World, viewed on http://projectdanube.org/wp-content/uploads/2011/10/The-Role-of-New-Media-for-the-Democratization-Processes-in-the-Arab-World.pdf, p. 13, retrieved on January 29th 2013.

[57] Freedom House: Freedom on the Net. Libya 2012, retrieved on 31st of January 2013.

country, often via bloggers based in the diaspora.[58] Especially people living in the East of the state were able to use internet access uploading videos and updating Twitter about what was going on.[59]

4.2 What Media can do and can not do

Can the events going on in the Arab world be called a Facebook revolution, was one on the questions asked not only by the "Time Magazine" or "The New York Times". *"While these revolutions may have occurred without social media, the speed at which they have occurred would have been much, much different"*[60], Robert McNab mentioned in a further movie made by NATO. But, it is clear that social media was quite important during the Arab uprisings. The anonymity of Facebook and Twitter can be seen as a safe place to hide, in a both positive and negative way. People uses media like Facebook for searching for information if a government provides Internet access. But what is different nowadays to revolutions in the end of 1980's and beginning of 1990's? In the past revolutions took place ruled by a leader or a cadre of leaders of rebel groups.[61] Today, it has been changed. The motivation to revolt can be found within individuals, loosely associated in a Facebook group. McNab: *"We see democratic change, or what we hope is democratic change occur in a very broad base of support that is leaderless."*[62] Fact is, social mass media was supporting the Arab awakening fundamentally. The Internet has the ability to maintain transparency or, at least, to show what is going right or wrong within the affected states or the international community. And: people started to express themselves and to get together in an association of other people thinking, feeling and expressing in the same way. The place where such a revolution starts was relocated to the notebook before. Afterwards, the relocated crowd started to come together on the local market square, with the difference to the Velvet Revolution 1989/1990: the get-together today was well organized, not merely by chance. People were informed how much protesters will get together and where. The level of security and protection was higher than in the past. More people do form a higher suspension regarding to the regime which want to intervene. All in all, the more people come together, the lower is the risk to be killed by armed

---

[58] Ib., retrieved on 31st of January 2013.
[59] Ib., retrieved on 31st of January 2013.
[60] NATO Review: Political change. What social media can - and can't do, viewed on http://www.nato.int/docu/ review/2011/Social_Medias/Social_media_can_do/EN/index.htm, retrieved on 29th of January 2013.
[61] Ib., retrieved on 29th of January 2013.
[62] Ib., retrieved on 29th of January 2013.

forces, the police or other supporter of the leading government. And people start to realize that they are not the only one who are going to change the circumstances.

But these are not the only advantages which can be found. The use of social mass media spread that quickly superficially because of the low costs, if it is even not totally free of charge. Facebook is free, Twitter is free, web space can be found for free and You Tube is free, too. *"Individuals can organise almost costlessly, share information almost costlessly, and that allows them to share ideas, to gather, to call for change in a very short order, whereas before it would have been much more difficult"*[63], is said in the movie. Social media has partly taken a big credit in the Arab awakening, but: *"For those people who call this the Twitter or Facebook Revolution I think that's a misnomer that is unfair to the people who went out on the streets and risked or even lost their lives in the cause of these revolutions."*[64] The role of social mass media seemed to be overestimated, as the facts tell. Social media found its limits during the course of the Arab revolution. In Egypt, for example, more and more people gather on the streets when the Internet access was stopped by the government. *"I think, social media starts the idea. It gets people involved. But then, the protest will take a life of its own"*[65], McNab mentioned. A German news magazine once wrote: *"People take to the streets to rise up against the status quo, they seek public exposure and they fight their battles using their bodies as their most important weapon. Without this physical presence, there can be no demonstrations – not even in the era of the disembodied Internet. But the protests also don't work without the Internet."*[66]

After all, let us summarize what media is able to do and not to do ideal-typically: a) political changing can be achieved faster and direct by networking people with the same interests and thoughts (intergroup relations), b) it allows a strategic real-time movement during a revolution, c) it can shorten revolutions and uprisings in comparison to revolts in the past, d) media can be seen as the nucleus of a safe get-together before people perform "offline" out on the streets (collective action), e) a typical single leader of a revolution was replaced by individuals with a wide variety of different actors (individual transformation), f) the mass media like television, radios and newspaper all around the world were affected by the pictures sent originally by the protesters being involved in that revolution (external attention), g) it can help to establish a transitioning system (regime policies). *"Using*

---

[63] Ib., retrieved on 29th of January 2013.

[64] Ib., retrieved on 29th of January 2013.

[65] Ib., retrieved on 29th of January 2013.

[66] Ulrike Knöfel: The Art of Demonstration. Exhibit Explores Emergence of Online Protests, in: Spiegel Online, viewed on http://www.spiegel.de/international/zeitgeist/the-art-of-demonstration-exhibit-explores-emergence-of-online-protests-a-809881.html, retrieved on 29th of January 2013.

connection technologies and social media networks, the people of all nations can interact with one another on the pressing issues of our time. They can share solutions to common problems, borrow and modify new ideas, invest in transnational social and commercial entrepreneurship, and entertain each other."[67]

On the other hand, mass media can be used as an instrument of propaganda, manipulation and infiltration by dictatorial leaders, regimes and governments to. The using of media does not mean that a) fully developed democracy will be established in the future. It neither can b) issue a guarantee that the aims of the people will be successfully achieved, nor c) that people are able to use the advantages of media in a meaningful way. „*Technology by itself is agnostic. It simply amplifies and extends the existing sociology in a community. If the community aspires to democracy, it facilitates that. If the community aspires to something else, it will facilitate that. And if it succeeds in toppling a power structure, there is no guarantee it will succeed in standing up a new one.*"[68]

That the use of media can be dangerous, too, it shows the issue of Libya. The government used the internet during the civil war to take control over the riots. Web pages were censored, and later the whole Internet access was shut down by the regime. *"The government also used the Internet to learn about the rebel movement's organizational structure and about individual actor's identities."*[69] Since the victory of the rebel groups in Libya in August 2011, the people were allowed to access all former blocked web pages again. *"As of May 2012, social media applications like the video-sharing website YouTube, the social-networking platform Facebook, and the microblogging service Twitter were freely accessible. YouTube had previously been blocked under the Qadhafi regime beginning in January 2010."*[70] During the conflict people living in the Eastern parts of Libya got access to YouTube again in the beginning of April 2011, in the West not before November 2011. Other options of social media like Facebook or Twitter were shut down by the regime for a couple of weeks in Februar 2011 shortly *"before the entire internet was cut off"*[71].

---

[67] NATO Review: Social media. Cause, effect and response, viewed on http://www.nato.int/docu/review/2011/ Social_Medias/21st-century-statecraft/EN/index.htm, retrieved on 29th of January 2013.

[68] Ib., retrieved on 29th of January 2013.

[69] Markus Sabadello: p. 6, retrieved on 29th of January 2013.

[70] Freedom House: Freedom on the Net. Libya 2012, retrieved on 31st of January 2013.

[71] Ib., retrieved on 31st of January 2013.

4.3 A possible Scenario

Would the protests probably have occurred in Libya without the help of Facebook or other social networks? The interdependency of Libya and the role of social mass media has to be deepen in this chapter. To generate a scenario which seems to be quite real, the author of this present essay have chosen the case when the protest wave of the Arab Spring spread over Libya on 15th of February 2011.

We are writing the year 2011. It is in the mid of February, day number 14 of the month, it is late in the evening. In Benghazi, a city with approximately 700,000 inhabitants, located in the North-East of the country, directly at the Mediterranean Sea, people are watching television. The atmosphere in the second largest city in a country with 5.6 million inhabitants consists of an embarrassing silence. The state television is screening a distorted propaganda movie made by the Libyan government about the revolts in Tunisia and Egypt. Gaddafi, which is in power in Libya as a dictatorial leader for almost 40 years, knows how to influence his folk by presenting controlled news in television, radio or newspapers. People have been influenced by that for many decades. They simply are not able to be reflective and open-minded. The people just forgot how to evaluate state-controlled news. Gaddafi is screening himself on the TV channels: *"The young people [in Tunisia and Egypt] at the core of the protest movement are acting under the influence of hallucinogenic drugs and the demonstrations are controlled by al-Qaeda."*[72] He is adding: *"The demonstrations are led by foreigners."*[73] But the fast growing population, especially the youth, are even better educated than before. They are mistrusting Gaddafi and his regime. One third of the population in Libya is under the age of 15, almost 30 percent are between 15 and 29 years of age and more than 20 percent are between 30 and 44 years old.[74] Generally the typical Internet and social media user is a young man or woman. Due to the high unemployment rate of 30 percent[75] in the country, which is especially affecting influencing young people, a high level of uprising potential can be found in this group. That burgeoning aggression touchable and visible during the night in Benghazi.

---

[72] Due to this scenario the author of the essay has changed the original statement. The original indirect quotation can be found here: Encyclopædia Britannica Online: Libya Revolt of 2011, retrieved on 30th of January 2013.

[73] Due to this scenario the author of the essay has changed the original statement. The original indirect quotation can be found here: Emad Mekay: Middle East. One Libyan Battle Is Fought in Social and News Media, in: New York Times of 23rd February 2011, viewed on http://www.nytimes.com/2011/02/24/world/middleeast/24iht-m24libya.html?_r=2&, retrieved on 30th of January.

[74] Encyclopædia Britannica Online: Libya Revolt of 2011, retrieved on 30th of January 2013.

[75] Ib., retrieved on 30th of January 2013.

The emotions sharpened within the next day, 15th of February 2011, when Fethi Tarbel, a human rights lawyer, was imprisoned by the police.[76] People have watched the performing of the police in publicity, directly on the streets. They have seen how brutal they have arrested Tarbel. And they have recorded the scene with their mobile phones. But the regime reacted soon, too, and they shut down all Internet accession in Libya. Only state television was still screening. And in the news the commentators have not lose a word about the things happened in Benghazi. The people simply did not knew what was going on outside. Instead of the revolts emerged afterwards in Libya in reality, in this possible scenario people had no chance to share their recordings via Internet or media. Due to the stop of all internet connections the people did not have the ability to spread information through the country. The revolution nipped in the bud for the moment. But due to the domino effect emerged in the countries around Libya the wave of protest would spread over the sate sooner or later, but the people easily can not count on the Internet as a supporter in getting organized during the revolution. It is resulting in a long-lasting civil war with an increased number of killed or injured citizens than in reality. The infrastructure in the whole country is getting completely destroyed and Gaddafi, due to an badly organized rebel group, is still in power. Nothing changed for the better in Libya so far if the Internet would be shut down by the government.

5. Abstract

In the beginning of this present essay the author gave some theoretical inputs regarding to the discussed topic. Later, these definitions were used to move forward to the practical framework. The situation behind the emergence of the Arab Spring generally was explained, as well as the further course of the revolts in Libya. The focus was in particular on the reasons of the revolution in Libya and on the results afterwards, too. It was complemented by interesting facts, quotations and statistics.

The second main aspect was focussed on the role of social mass media with concentration on advantages, disadvantages and limits of media regarding to the revolution in the Arab world. "What media can do and can not do?" was one of the research questions which was answered here. In the last chapter of the main part the author was focussed on a construction of a imaginary scenario which might come true when media would be switched of by the former regime in Libya.

---

[76] Ib., retrieved on 30th of January 2013.

What we have learned during this essay is that social mass media and the Arab Spring is connected towards each other, but not to the extent how Western media is dealing about that. The interdependency between media and the uprisings in the Arab world is given, but only weakly tied. As it was said quite often within the interviews mentioned, media was important to support the emergence of the protests. It was used to spread "bad" news over the country and to share it with people thinking the same. Furthermore, it was used to organize the revolts in the internet. But this is only one part of the Arab Spring. The revolution would not have happened without people going out on the streets and fighting for their rights to have a better life. Both aspects have to be recognized as an important element of the course of the Arab Spring.

All in all, the Arab Spring should neither be called "Facebook revolution" nor "Twitter revolution", even though a digital movement has been included. Especially the *"loose and non-hierarchical organisation of the protest movements unconsciously modelled on the networks of the web"*[77] is a new development in comparison to former uprisings. Though, the *"importance and impact of social media on each of the rebellions we have seen [...] has been defined by specific local factors"*[78].

In an article written Markus Sabadello is said: *"A more accurate statement might be that Youtube, Facebook, Twitter & Co. merely constitute new weapons for both sides of an old fight."*[79] And: *"New media seem to offer development of new forms of political activism beyond the now old 'new social movement' type."*[80]

Besides of the using of media, Libya is still confronted with an unclear future. The new government has to lead the whole nation, the policy, the economy, associations, the military and the people through a long-lasting transition. And the Libyans can not be sure if the transition of their country will be a successful one. But, what they can support is to work on the emergence of a strong civic society, respecting democratic principles, political parties and other associations of people. Media can still support the success of this transition. The Libyan society is suffering from a fractured being, fragmented by different cleavages. *"This lack of social and governmental cohesion will hamper any prospective transition to democracy. Libya must first restore security and introduce the law and order missing for*

---

[77] Peter Beaumont: Can social networking overthrow a government?, in: The Sydney Morning Herald of 25th February 2011, viewed on http://www.smh.com.au/technology/technology-news/can-social-networking-overthrow-a-government-20110225-1b7u6.html, retrieved on 30th of January 2013.
[78] Ib., retrieved on 30th of January 2013.
[79] Markus Sabadello: p. 13, retrieved on 30th of January 2013.
[80] Annabelle Sreberny: p. 4, retrieved on 30th of January 2013.

decades under Qaddafi's regime."[81] It is the new government's both turn and challenge to reconstruct infrastructure and administration, to produce trust between the citizens and the government, to open media and to be opened for non-governmental organizations as a fundament for a functional democracy. *"While Tunisia and Egypt grapple in their own ways with building political institutions – constitutions, political parties, and electoral systems – Libya will need to begin by constructing the rudiments of a civil society."*[82] One can, in the end, only hope that the road that lies ahead of Libya is one that leads to a healthy democracy that aims towards the improvement of the living conditions of the people who live there.

6. Bibliography

Al Jazeera: Taking power through technology in the Arab Spring, viewed on http://www.aljazeera.com/indepth/opinion/2012/09/2012919115344299848.html, retrieved on 24th of January 2013.

Alex Comninos: Twitter revolutions and cyber crackdowns. User-generated content and social networking in the Arab spring and beyond, 2011.

Anja Türkan: Digital Diplomacy. Der Wandel der Außenpolitik im digitalen Zeitalter, Stuttgart 2012.

Annabelle Sreberny: New Media and the Middle East. Thinking allowed, in: University of Michigan: Journal, No. 2 (Spring 2012), viewed on: http://www.lsa.umich.edu/UMICH/ii/Home/II%20Journal/Documents/2012spring_iijournal_a rticle1_sreberny.pdf, retrieved on 30th of January 2013.

Annabelle Sreberny: New Media and the Middle East. Thinking allowed, in: University of Michigan: Journal, No. 2 (Spring 2012), viewed on: http://www.lsa.umich.edu/UMICH/ii/Home/II%20Journal/Documents/2012spring_iijournal_a rticle1_sreberny.pdf, retrieved on 29th of January 2013.

---

[81] Lisa Anderson: Demystifying the Arab Spring. Parsing the Differences Between Tunisia, Egypt, and Libya, in: Foreign Affairs, Volume 90, No. 3, viewed on http://www.ssrresourcecentre.org/wp-content/uploads/2011/06/ Anderson-Demystifying-the-Arab-Spring.pdf, p. 7, retrieved on 30th of January 2013.
[82] Ib., p. 7, retrieved on 30th of January 2013.

Bibliographisches Institut & F.A. Brockhaus AG: Medien, viewed on http://www.brockhausenzyklopaedie.de/be21_article.php?document_id=0x0919d990@be, retrieved on 26th of January 2013.

Countries and their political status being affected by the Arab Spring, viewed on https://moodle.hampshire.edu/pluginfile.php/69048/course/section/31642/arab-spring-map1.jpg, retrieved on 31st of January 2013.

Echo Keif: We Are All Khaled Said. Revolution and the Role of Social Media, Fairfax 2011.

Emad Mekay: Middle East. One Libyan Battle Is Fought in Social and News Media, in: New York Times of 23rd February 2011, viewed on http://www.nytimes.com/2011/02/24/world/middleeast/24iht-m24libya.html?_r=2&, retrieved on 30th of January.

Encyclopædia Britannica Online: Arab Spring, viewed on http://www.britannica.com/EBchecked/topic/1784922/Arab-Spring, retrieved on 26th of January 2013.

Encyclopædia Britannica Online: Communication, viewed on http://www.britannica.com/EBchecked/topic/129024/communication, retrieved on 26th of January 2013.

Encyclopædia Britannica Online: Libya Revolt of 2011, viewed on http://www.britannica.com/EBchecked/topic/1766291/Libya-Revolt-of-2011, retrieved on 30th of January 2013.

Encyclopædia Britannica Online: Public Opinion. The mass media, viewed on http://www.britannica.com/EBchecked/topic/482436/public-opinion/258760/The-mass-media#ref397897, retrieved on 26th of January 2013.

Encyclopædia Britannica Online: Social Network, viewed on http://www.britannica.com/EBchecked/topic/1335211/social-network, retrieved on 26th of January 2013.

Facebook: We are all Khaled Said, viewed on http://www.facebook.com/elshaheeed.co.uk?fref=ts, retrieved on 29th of January 2013.

Freedom House: Freedom of the Press. Libya 2012, viewed on http://www.freedomhouse.org/report/freedom-press/2012/libya, retrieved on 31st of January 2013.

Freedom House: Freedom on the Net 2012. Global Scores, viewed on http://www.freedomhouse.org/sites/default/files/resources/FOTN%202012%20-%20Tables%20and%20Charts%20FINAL.pdf, retrieved on 31st of January 2013.

Freedom House: Freedom on the Net. Libya 2012, viewed on http://www.freedomhouse.org/report/freedom-net/2012/libya, retrieved on 31st of January 2013.

Lina Ben Mhenni: Vernetzt euch!, Berlin 2011.

Lisa Anderson: Demystifying the Arab Spring. Parsing the Differences Between Tunisia, Egypt, and Libya, in: Foreign Affairs, Volume 90, No. 3, viewed on http://www.ssrresourcecentre.org/wp-content/uploads/2011/06/Anderson-Demystifying-the-Arab-Spring.pdf, retrieved on 30th of January 2013.

Markus Sabadello: The Role of New Media for the Democratization Processes in the Arab World, viewed on http://projectdanube.org/wp-content/uploads/2011/10/The-Role-of-New-Media-for-the-Democratization-Processes-in-the-Arab-World.pdf, retrieved on 30th of January 2013.

NATO Review: Arab Spring = Facebook revolution #1?, viewed on http://www.nato.int/docu/review/2011/Social_Medias/Arab_Spring/EN/index.htm, retrieved on 29th of January 2013.

NATO Review: Political change. What social media can – and can't do, viewed on http://www.nato.int/docu/review/2011/Social_Medias/Social_media_can_do/EN/index.htm, retrieved on 29th of January 2013.

NATO Review: Social media. Cause, effect and response, viewed on http://www.nato.int/docu/review/2011/Social_Medias/21st-century-statecraft/EN/index.htm, retrieved on 29th of January 2013.

NATO Review: The reconstruction will not be tweeted, viewed on http://www.nato.int/docu/review/2011/Social_Medias/Social_Media_changes/EN/index.htm, retrieved on 29th of January 2013.

Peter Beaumont: Can social networking overthrow a government?, in: The Sydney Morning Herald of 25th February 2011, viewed on http://www.smh.com.au/technology/technology-news/can-social-networking-overthrow-a-government-20110225-1b7u6.html, retrieved on 30th of January 2013.

Small Wars Journal: Social Media and the Arab Spring, viewed on http://smallwarsjournal.com/jrnl/art/social-media-and-the-arab-spring, retrieved on 24th of January 2013.

Twitter Revolution. How the Arab Spring was helped by Social Media, viewed on http://www.policymic.com/articles/10642/twitter-revolution-how-the-arab-spring-was-helped-by-social-media, retrieved on 24th of January 2013.

Ulrike Knöfel: The Art of Demonstration. Exhibit Explores Emergence of Online Protests, in: Spiegel Online, viewed on http://www.spiegel.de/international/zeitgeist/the-art-of-demonstration-exhibit-explores-emergence-of-online-protests-a-809881.html, retrieved on 29th of January 2013.

Wael Ghonim: Revolution 2.0. The Power of the People Is Greater Than the People in Power. A Memoir, New York 2012.